Procedures in FIELD GEOLOGY

Tom Freeman
Distinguished Teaching Professor Emeritus
University of Missouri-Columbia

b
Blackwell
Science

© 1999 by Tom Freeman

BLACKWELL PUBLISHING
350 Main Street, Malden, MA 02148-5020, USA
9600 Garsington Road, Oxford OX4 2DQ, UK
550 Swanston Street, Carlton, Victoria 3053,
   Australia

The right of Tom Freeman to be identified as the
Author of this Work has been asserted in
accordance with the UK Copyright, Designs, and
Patents Act 1988.

All rights reserved. No part of this publication may
be reproduced, stored in a retrieval system, or
transmitted, in any form or by any means,
electronic, mechanical, photocopying, recording or
otherwise, except as permitted by the UK
Copyright, Designs, and Patents Act 1988, without
the prior permission of the publisher.

First published 1999 by Blackwell Science Ltd,
a Blackwell Publishing company

5   2006

ISBN-13: 978-0-86542-008-3
ISBN-10: 0-86542-008-4

Printed and bound in the United Kingdom
by Athenæum Press Ltd, Gateshead, Tyne & Wear

The publisher's policy is to use permanent paper
from mills that operate a sustainable forestry
policy, and which has been manufactured from
pulp processed using acid-free and elementary
chlorine-free practices. Furthermore, the publisher
ensures that the text paper and cover board used
have met acceptable environmental accreditation
standards.

For further information on
Blackwell Publishing, visit our website:
www.blackwellpublishing.com

Preface

My purpose in writing this manual is to provide a pocket-size presentation of field procedures without the bulk and cost of a comprehensive textbook.

The first part, *Tools of the Trade*, covers the Brunton compass and Jacob's staff, the Silva compass, and the plane table and alidade—along with principles of map direction and strike and dip. Also, I have included a section on uses of the stereographic net for those who lack hardware and software necessary for the manipulation of field measurements.

A second part, *Things to Do*, presents a variety of procedures commonly undertaken in geologic field courses.

The third part, *Reference Stuff*, provides basic information on topographic maps, trigonometric solutions, and conventional map patterns and symbols.

Field procedures described herein are those that have emerged through years of teaching at our University of Missouri's Camp E.B. Branson in the Wind River Mountains of Wyoming. I welcome suggestions for the inclusion of additional procedures in future editions.

Tom Freeman
Columbia, Missouri

CONTENTS

TOOLS OF THE TRADE

The Brunton compass *1*
   But first—two methods of stating direction *1*
      Bearing (or quadrant) method *1*
      Azimuth method *2*
   Brunton anatomy *3*
   Magnetic declination *4*
      Magnetic declination defined *4*
      Adjusting a Brunton for magnetic declination *7*
      Beware of metal objects! *7*
   Measuring direction to an object *8*
   Using a Brunton as a protractor *11*
   Measuring vertical angles *12*
   Review—trigonometry of a right-triangle *14*
   Solving for map distance represented by slope angle and slope distance *15*
   Solving for difference in elevation represented by slope angle and slope distance *15*
   Solving for difference in elevation using succesive eye-height measurements *16*
   Strike and dip—definitions *17*
   Recording strike with the right-hand rule *18*
   The other right-hand rule *18*
   The azimuth method of describing the orientation of and inclined plane *19*
   Measuring strike—contact method *20*
   Measuring dip magnitude—contact method *21*
   Using two outcrops to measure strike and dip *22*
   Measuring trend and plunge of a lineation *24*
   Measuring the trace of an outcrop *25*
   Measuring an inclined stratigraphic section with a Jacob's staff and Brunton *26*

The Silva compass 28
- Weaknesses and strengths 28
- Silva anatomy 29
- Setting magnetic declination on a Silva 30
- Measuring direction with a Silva 30
- Measuring strike and dip—contact method 31
- Plotting directions on a map with a Silva 32
- Using a Silva to measure rake (or pitch) 34

The plane table and alidade 36
- Beaman scales, type 1 and type 2 36
- Alidade anatomy 37
- Setting up the table and working on a base map 38
- Orienting the table, focusing, adjustments 39
- Reading the vertical arc and vernier 40
- Stadia distance 41
- Procedure—'running the gun' 42
- Computing elevationns—forsights 44
- Computing elevations—backsights 45
- A sample format for recording data 46

The stereographic net 48
- Purpose 48
- Preparing a stereographic net 50
- Planar and linear features 51
- Representing an inclined (dipping) plane 51
- Representing an inclined (plunging) line 53
- Representing the plunge of a line from its rake in an inclined plane 54
- Representing a line formed by the intersection of two planes 56
- Solving for original orientation of structurally tilted cross-beds 57
- Solving for strike and true dip with two apparent dips 61

THINGS TO DO

Confronting an outcrop—points to ponder 62
  The value of reconnaissance 62
  The utility of field sketches 62
  The importance of contact relationships 63
Mapping—pace-and-Brunton method 64
  Definition and procedure 64
  Correcting error of closure 66
Triangulating—intersection method 68
  Resection method 69
Describing stratigraphic sections—format 70
  Lithic descriptions 70
  Geomorphic profile 71
Constructing a geologic road traverse 72
Using contour maps—the 'rules of contours' 74
Constructing a contour map 76
Constructing a topographic profile 78
Constructing a geologic cross-section 80
Solving for exaggerated dip when the vertical scale is exaggerated 82
Solving for true dip when a cross-section is not perpendicular to strike 83

REFERENCE STUFF

Land survey system—latitude and longitude 84
Townships and ranges 85
Sections and 86
Map dimensions and scales 88
Grid north 91
Trigonometric solutions of stratigraphic thickness 92
  Where slope and dip are in opposite directions 92
  Where slope and dip are in the same direction 93
Lithic patterns and symbols 94
Structural map symbols 95

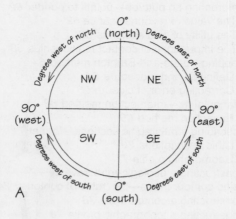
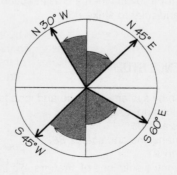

Fig. 1. The bearing (or quadrant) method of stating direction. (A) Four quadrants—NE, SE, SW, NW—are east or west of north or south. (B) The bearing is the angle turned east or west from north or south.

THE BRUNTON COMPASS

But first...two methods of stating direction

The bearing (or quadrant) method—is no doubt a carry-over from the ancient compass rose with its four quadrants. A bearing is an angle measured eastward or westward from either north or south, whichever is closer. The method employs a circle divided into four quadrants: northeast (NE), northwest (NW), southeast (SE), and southwest (SW) (Fig. 1A). Each of the four quadrants is divided into 90°, beginning with 0° at the north and south poles and ending with 90° at east and west. So, bearing is always less than 90° measured eastward or westward from either the north pole or the south pole.

A bearing direction can be specified by stating (first) the pole—north or south—from which the angle is measured; (second) the magnitude of the angle measured; and (third) the direction—east or west—toward which the angle is measured. Four examples are shown in figure 1B.

By the way...if you compare the face of a compass with figure 1A, the compass might appear to be mis-labeled (i.e., west in place of east, and east in place of west). A compass is labeled in this manner so that when you rotate it progressively westward (for example) the compass needle "reads" progressively westward (rather than progressively eastward).

The azimuth method—of stating direction employs a circle divided into 360°, beginning with 0° at the north pole and increasing clockwise to 360° at the north pole (i.e., 0° and 360° are coincident) (Fig. 2). An azimuth circle is graduated in a manner analogous to that of the face of a clock. Only instead of being a clockwise sweep of 60 minutes, an azimuth circle is a clockwise sweep of 360 degrees.

The four directions illustrated in figure 2 are the same as those illustrated in figure 1, which serves to contrast one method with the other.

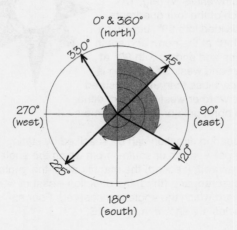

Fig. 2. The azimuth method of stating direction.

The Brunton compass is available with either the quadrant or the azimuth circle. The quadrant circle is more traditional, but the azimuth circle is less subject to recording-error; and, azimuth data are more easily processed with a computer. The Brunton in figure 3 has an azimuth circle.

BRUNTON ANATOMY

The Brunton compass (Fig. 3)—patented by D.W. Brunton in 1894—is the Swiss Army knife of field geologists.

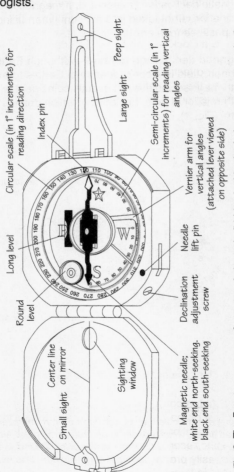

Fig. 3. The Brunton compass. A quarter-circle scale (hidden in this view) shows percent-grade. A scale on the vernier arm provides for reading fractions of one degree when mounted on a tripod.

MAGNETIC DECLINATION

When using a Brunton (or any other magnetic device) to measure direction, an understanding of magnetic declination is essential. If magnetic declination is not taken into account when using a compass, serious errors can result.

Magnetic declination defined—Although Earth's magnetic field is a consequence of Earth's rotation, north magnetic pole is not exactly coincident with north rotational pole, the latter of which defines 'true north' (Fig. 4).

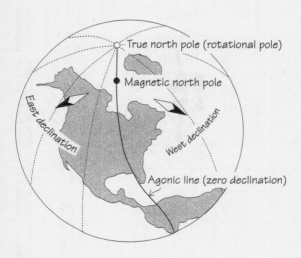

Fig. 4. White arrowheads that point toward Earth's magnetic north pole diverge (or decline) from meridians (dotted lines), which converge on 'true north' pole. Along the agonic line, directions to the two poles are the same. That is, along the agonic line a magnetic needle is parallel to a meridian.

Some map legends schematically indicate both the *direction* (i.e., east or west of true north) and the *amount* of magnetic declination applicable to the mapped area (Fig. 5). (Grid north —GN in figure 5—is discussed on page 91.)

Fig. 5. Map margin information showing directions of grid north (GN), true north (star), and magnetic north (MN). In this case grid north is west of true north by zero degrees, 57 minutes, which is equal to 17 mils. (There are 6,400 mils in a circle.) Magnetic north is east of true north by 8 degrees, which is equal to 142 mils. Notice that the angles illustrated are nowhere near their actual stated values. Lines GN and MN are schematically drawn only to indicate whether they're east or west of true north.

Figure 6 shows differences among magnetic declination values within the United States, along with their annual rates of change. Because of local variations in Earth's magnetic field, declination lines are irregular. A magnetic needle does not point exactly toward Earth's magnetic pole (surprise!), but it could lead you there.

If you ever need to know the exact magnetic declination of a particular place at a particular time, you can phone the offices of the United States Geological Survey in Rolla, Missouri, to inquire. (Phone 573/341-0998.)

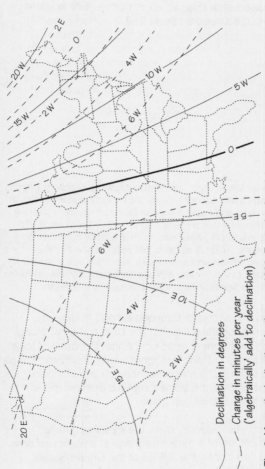

Fig. 6. Magnetic declination for the year 1990. Example of applying change in minutes per year: In the year 2000 Seattle will have a declination of 19 degrees E (i.e., 10 x 6 minutes W, added to 20 degrees E).

Adjusting a Brunton for magnetic declination—Using the declination-adjustment screw on the side of the Brunton (Fig. 3) turn the graduated circle relative to the fixed index pin an amount equal to the magnetic declination. The question arises: In case of east magnetic declination—for example—should the graduated circle be turned clockwise, or should it be turned counter-clockwise? The answer is...*clockwise*. (Conversely, if magnetic declination is west, you must turn the graduated circle counter-clockwise.)

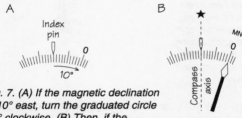

Fig. 7. (A) If the magnetic declination is 10° east, turn the graduated circle 10° clockwise. (B) Then, if the compass is rotated so that it reads due north (by bringing the north-magnetic-seeking needle to zero degrees), the axis of the compass aligns with true north.

Beware of metal objects!—Metal objects, except those made of brass, will deflect a compass needle (e.g., your belt buckle, mechanical pencil, knife, and hand-lens). A battery-powered calculator can deflect a needle 6° or more.

A compass needle near a power line will tend to be deflected in a direction perpendicular to the line.

MEASURING DIRECTION TO AN OBJECT

Case I—When the elevation of an object is not more than a few tens of degrees above—nor more than approximately 15° below—the elevation of the viewer, a Brunton should be held as in figure 8.

Fig. 8. How to hold a Brunton when measuring the direction to an object whose elevation is not more than a few tens of degrees above—nor more than approximtely 15° below—the elevation of the viewer.

Step 1—With your elbows hugging your sides for support, hold the Brunton approximately level in one hand (as indicated by the round level). With your free hand, swing the large sight and the mirror about their hinges to positions that allow you to view the reflection of the object either through the slot in the large sight or along its tip. (A little trial-and-error is required here.) Then bring the round level to center. You will no doubt have to pivot your body and Brunton a bit to maintain your view of the object along the axis of the large sight. You must align your eye, the axis of the large sight, the center line of the mirror, and the object. You may move your head in order to achieve this alignment, so long as the round level remains centered.

Step 2—Read the bearing or azimuth indicated by the white (north-seeking) end of the needle.

USING A BRUNTON A A PROTRACTOR

If you find yourself without a protractor for plotting rays on a map in the field—as you must do when triangulating—you can first field-orient your map and then use the Brunton as a protractor:

Step 1—(a) Place the map on level ground (best on a map board), (b) place the edge of the Brunton along a north-south section line, and (c) rotate the map to bring the Brunton's needle to zero index. The map is now field-oriented. (Be sure that you're not 180° in error!)

With the map in this position, you can construct a line on the map representing your line of sight to a feature (whose bearing or azimuth you have measured) by either step 2A or step 2B (below), depending on the purpose of the line.

Step 2A—To draw a line from a feature shown on the map toward your position not shown: (a) Place the edge of the Brunton at that feature, (b) rotate the Brunton about that feature to a position where the needle indicates the feature's bearing or azimuth, and (c) draw a line along the straight edge of the Brunton from the feature toward you.

Step 2B—To draw a line from your position shown on the map toward a feature not shown: Follow step 2A, but place the edge of the Brunton at your position, rotate as in step 2A, and draw a line from your position toward the feature.

MEASURING VERTICAL ANGLES

The long level, together with attached vernier arm and companion semi-circular scales, enables a Brunton to function as a clinometer (Fig. 11). The lever on the Brunton's base turns the long level and attached vernier.

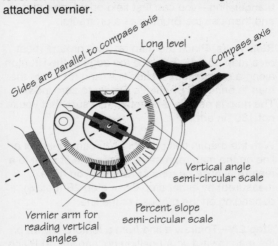

Fig. 11. A Brunton positioned as a clinometer. The inner larger semi-circular scale measures vertical angles. The outer smaller semi-circular scale measures the percent grade of a slope.

As a clinometer, a Brunton can be used to measure the inclination of a surface on which it rests. Or, it can be hand-held and used to measure the angle of a slope (Fig. 12). When measuring slope angle…

Step 1—Fold the hinged peep-sight tip of the large sight so that you can sight either through the peep-sight or along its point, through the sighting window of the mirror, to the object (Fig. 12A).

Step 2—Swing the lid into a position that allows viewing the reflection of the compass face.

Step 3—Rotate the long level (using its lever on the base of the Brunton) to level position.

Step 4—Look again at the object and adjust the long level if needed. Repeat as necessary.

Step 5—Remove your fingers from the long-level lever, look directly at the face of the compass, and read the angle or percent slope.

When working with a field partner (Fig. 12B), each partner should voice his/her individual reading. If different, average the two readings.

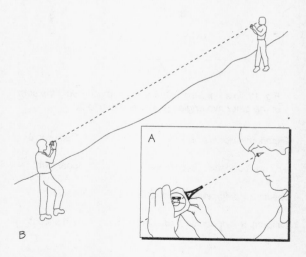

Fig. 12. Measuring slope angle with a Brunton. Note: The farther the Brunton is held away from the eye, the more accurate the measurement.

REVIEW—TRIGONOMETRY OF A RIGHT-TRIANGLE

The following (Fig. 13 and below) is for convenient reference when solving for horizontal distance and vertical distance as elements of a right-triangle.

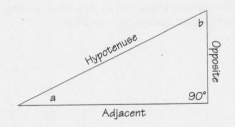

Fig. 13. In a right-triangle (i.e., one angle = 90°) the sum of the other two angles (a + b) is equal to 90°.

Trigonometric relationship	Reciprocal
sine a = opposite / hypotenuse	cosecant
cosine a = adjacent / hypotenuse	secant
tangent a = opposite / adjacent	cotangent

and…$(\sin a)^2 + (\cos a)^2 = 1$

and…$\sin a / \cos a = \tan a$

SOLVING FOR MAP DISTANCE REPRESENTED BY SLOPE ANGLE AND SLOPE DISTANCE

Refer to figure 14. After determining slope angle (y)—as in figure 12—tape or pace to determine slope distance (S), then solve for the horizontal distance (H) by...

$$H = S \cos y$$

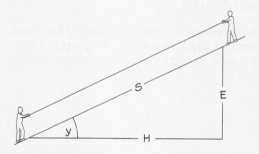

Fig. 14. Measuring slope distance (S) with a tape. Slope angle (y), elevation difference (E), and horizontal difference (H) are also labeled.

SOLVING FOR DIFFERENCE IN ELEVATION REPRESENTED BY SLOPE ANGLE AND SLOPE DISTANCE

Refer to figure 14 once again. Solve for the difference in elevation (E) by...

$$E = S \sin y$$

SOLVING FOR DIFFERENCE IN ELEVATION USING SUCCESSIVE EYE-HEIGHT MEASUREMENTS

Set the long level of the Brunton at zero index and hold it as if measuring a vertical (i.e., slope) angle (see again figure 12). Or, more conveniently, use a tube-like instrument called a hand-level that is made for the purpose. Spot a point up-slope that is the same elevation as your eye, move to that spot and repeat the procedure until you arrive at your destination (Fig. 15).

The difference in elevation between point A and point B in figure 15 is 4 x eye-height plus the estimated distance between eye-height #4 and point B.

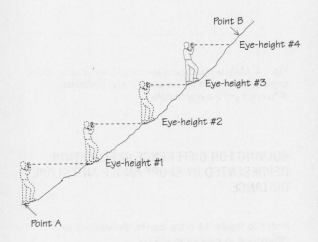

Fig. 15. Determining difference in elevation by using a Brunton as a hand-level.

STRIKE AND DIP—DEFINITIONS

Strike—is the map direction of the line formed by the intersection of (a) an inclined plane and (b) an imaginary horizontal surface. The top of the block diagram in figure 16 represents an imaginary horizontal surface, and the shaded layer represents an inclined limb of a fold bounded by planes. Their line of intersection is strike, which, in this case, is oriented north-south. Strike has traditionally been expressed in terms of the acute angle between the line of intersection and north, for example...N. 30° E. (instead of S. 30° W. nor E. 60° N.).

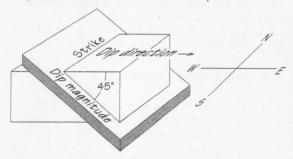

Fig. 16. Illustrating strike and dip of an inclined plane (e.g., that of a limb of a fold, or that of a fault). Strike is due north; dip magnitude is 45°; dip direction is east.

Dip—has two attributes: magnitude and direction. Dip magnitude is the (vertical) angle of inclination of the inclined plane. Dip direction is simply the direction toward which the plane is inclined downward (i.e., its 'down-hill' direction). Because dip magnitude is the maximum angle formed between the inclined plane and the imaginary horizontal surface, dip direction is necessarily perpendicular to strike (Fig. 16), so only the general direction of dip need be given (e.g., NW, SE, etc.).

RECORDING STRIKE WITH THE RIGHT-HAND RULE

A growing practice is to express strike simply as an azimuth value. But, inasmuch as a line (in this case, a strike line) has two directions, a choice must be made. By convention, American geologists record the azimuth along which one looks while positioned so that the dip direction is to his/her right—hence, the 'right-hand rule.' The person in figure 19 is positioned so that he can apply the right-hand rule. Dip direction is to his right; and, he has the Brunton oriented with the large sight away from him so that he can read the strike azimuth directly with the white (north-seeking) end of the needle. Only the magnitude of dip need be recorded, inasmuch as the direction of dip is indicated by the strike value (i.e., dip direction has to be to the right of strike).

THE OTHER RIGHT-HAND RULE

The right-hand rule described above might better be called the 'right-arm rule'—to distinguish it from the other right-hand rule used by British geologists. To apply the other right-hand rule (Fig. 17): Position yourself so that the thumb of your right hand points down-dip while the heal of your hand is flat on the bed's surface. Your index finger is extended to point in the direction of the recorded strike. Result: An implied dip direction opposite that derived from the 'right-arm rule' above.

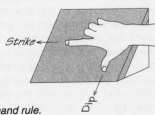

Fig. 17. The other right-hand rule.

THE AZIMUTH METHOD OF DESCRIBING THE ORIENTATION OF AN INCLINED PLANE

The idea of strike and dip must have been conceived in the dark of the moon. A much more practical method for describing the orientation of an inclined plane is to simply state the azimuth and magnitude of its dip. (Strike is inferred as 90° from the azimuth of dip—in case you need to know it.) For example, an inclined plane whose orientation might be described as...

Strike: N 45° E, dip to the southeast at 30° (Fig. 18)

...can be more simply described as 135° at 30°.

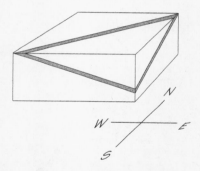

Fig. 18. Strike N 45° E, dip to the southeast at 30° can be more simply described as dip (azimuth) 135° at (magnitude) 30°.

The azimuth method is more direct, and azimuth is more easily digitally processed (in computers) than is strike. The stereographic-net solution of the original orientation of structurally-tilted cross-beds (page 57) uses this method to good advantage.

MEASURING STRIKE—CONTACT METHOD

Few beds are smooth enough to allow for making 'contact' strike and dip measurements. However, a bed's irregularities can be 'smoothed' by placing a map-board on the bed's surface (Fig. 19). Then, when the bottom side-edge of a Brunton is held against the inclined map-board and the round bubble is centered, the axis of the Brunton is parallel to strike, so you can read its bearing or azimuth.

Fig. 19. The contact method of measuring strike.

MEASURING DIP MAGNITUDE—CONTACT METHOD

The contact method of measuring dip magnitude can be achieved by using a Brunton as a clinometer (Fig. 20). Turn the Brunton on edge against the inclined surface and swing it to the position of maximum vertical angle indicated by the long level. A bit of trial-and-error is required to find the maximum angle.

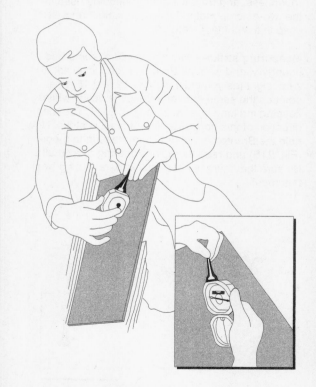

Fig. 20. The contact method of measuring dip magnitude.

USING TWO OUTCROPS TO MEASURE STRIKE AND DIP

A most accurate method—Two outcrops of the same layer on opposite sides of a valley present the most accurate method for measuring strike and dip (Fig. 21). This method might not seem as tangible as the contact method, but it has two advantages: (a) the two-outcrop method avoids effects of bed roughness; and (b) the extent of bedding used in the two-outcrop method is larger-scale than that used in the contact method.

Measuring strike—With your head at some bed boundary, and with the Brunton's long level set at zero, sight (as you did in figures 12 and 15) to a point on the same marker across the valley while keeping the long level at zero (Fig. 21A). The map direction of this horizontal line-of-sight is strike. Now hold the Brunton in the direction-measuring mode (Fig. 21B) and read the point's bearing or azimuth. Be sure that there is not a fault separating the two outcrops!

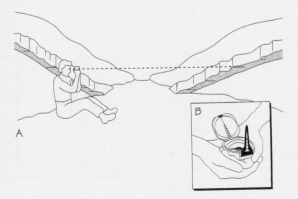

Fig. 21. The two-outcrop method for measuring strike.

Measuring dip magnitude—Hold the Brunton at arm's length and bring its edge into alignment with the stratigraphic marker sighted in figure 21 (Fig. 22). While holding it in this position, adjust the long level to its level position (i.e., center its bubble). Read the dip magnitude.

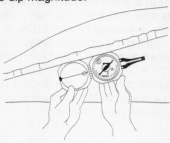

Fig. 22. Two-outcrop method for measuring dip magnitude.

Using a single outcrop—The technique illustrated in figures 21 and 22 can be applied to a single outcrop provided that there is enough bedding surface exposed to allow you to look along that surface. That is, if you can position yourself so that your line-of-sight is both (a) within the plane of the bed, and (b) horizontal, then your line of sight is strike. Dip can be measured as in figure 22.

Before closing, a 'trick of the trade'—The contact method of measuring strike won't work in cases of very low dip because the raised ring protecting the long-level lever prevents aligning the edge of the Brunton with the inclined surface. In such cases, measuring strike is still possible: Set the long level at zero and place the base of the Brunton flat on the inclined surface. Now rotate the Brunton as needed to bring the bubble of the long level to center. Can you see that the axis of the Brunton is now parallel to strike? Read the bearing or azimuth.

MEASURING TREND AND PLUNGE OF A LINEATION

Definitions—The trend of a lineation is the map direction (bearing or azimuth) in which it points downward. Its plunge is the vertical angle between it and a horizontal plane (i.e., its 'dip magnitude'). It is commonly difficult to apply contact methods to measuring the orientation of a lineation. Where this is the case, reasonably accurate measurements can be obtained as follows (Fig. 23 and steps 1–3):

Step 1—Position a pencil on the outcrop so that it is parallel to a lineation. (A bit of tape might be needed to hold the pencil in place.)

Fig. 23. Measuring the trend of a pencil positioned parallel to a lineation.

Step 2—Move to a position where you can look directly down the pencil. Hold the Brunton as in figure 23 (see again figure 9) and read the bearing (or azimuth) of the pencil's trend.

Step 3—While in the same position, hold the Brunton as in measuring a slope angle (see again figure 12A) and, while sighting down the length of the pencil, read its plunge.

MEASURING THE TRACE OF AN OUTCROP

A vein commonly occurs as a path of rubble with no hint of tabular shape (Fig. 24), so dip magnitude and direction are enigmatic. And, if the ground is inclined, strike is difficult to assess as well. Still, its trace (i.e., the direction of its outcrop) can provide important information in mineral exploration and structural studies. Of course, if you can trace the vein into an area of relief, you can there solve for its strike and dip.

Fig. 24. Sighting on your field partner is easier than sighting on a path of rubble. Average your two readings.

MEASURING AN INCLINED STRATIGRAPHIC SECTION WITH A JACOB'S STAFF AND BRUNTON

A Jacob's staff and a Brunton compass can be used to measure the stratigraphic thickness of a dipping stratigraphic section (Fig. 25).

Equipment—A Jacob's staff can be fashioned from a five-foot length of one-by-two lumber. The end that supports the Brunton must be sawed at a 90° angle with the length of the board so that the axis of the Brunton will be perpendicular to the staff.

Note: Your traverse must be in the direction of dip.

Step 1—Set the dip magnitude on the Brunton (Fig. 26), place the Brunton on the end of the staff, and lean it toward the section through an angle required to bring the long level to horizontal. The point sighted on the ground is five feet higher, stratigraphically, than the point at which you are standing.

Step 2—Go to the point sighted on the ground and repeat the procedure.

Step 3—Always measure in Jacob's-staff increments. Stratigraphic markers (e.g., lithic changes, unconformities, etc.) can be interpolated (by estimated footages) into the measured section.

Lateral off-setting—can be done freely in order to gain a topographic position that is more favorable for continuing your traverse. Simply follow a stratigraphic marker to its occurrence elsewhere, and resume measuring. It doesn't matter if you cross a fault in your off-set; but, if dip changes across a fault, you must re-set the dip magnitude.

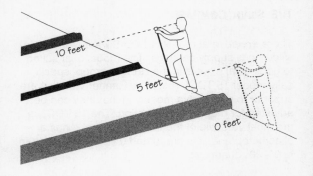

Fig. 25. Measuring the thickness of a dipping stratigraphic section with a 5-foot Jacob's staff and Brunton.

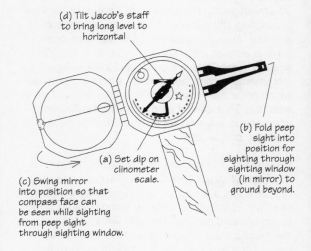

Fig. 26. Preparation of a Brunton on a Jacob's staff for measuring the thickness of a dipping stratigraphic section.

THE SILVA COMPASS

There are several models of the Silva advertised in wilderness supply catalogs. The model best suited to geological work is in the Ranger group, specifically model 15 TDCL (Fig. 27). Ranger varieties are superior to other Silva models in that they can be adjusted for magnetic declination. The 15 TDCL is the most versatile of the Rangers in that it has a clinometer; and, it is graduated 0°–360°, rather than in 0°–90° quadrants.

Weaknesses—The Silva compass is not designed for sighting vertical angles, so it cannot be used for measuring slope angles—as can the Brunton in figure 12—nor can it be used with a Jacob's staff as in figure 25. The Silva does have a clinometer that allows for the contact method of measuring dip; but, inasmuch as the Silva does not have a round level, measuring strike is not as accurate as with the Brunton. Also, because the compass face of the Silva is smaller than that of the Brunton, angular scales on the Silva are graduated in two-degree increments, rather than in one-degree increments, which limits the accuracy of the Silva still further. In sum, the Silva cannot compare with the Brunton compass as a surveying instrument. The Silva is crude by comparison.

Strengths—For plotting field measurements on a map, the Silva is superior to the Brunton because the Silva has a broad flat base (graduated both in centimeters/millimeters and in inches) that is good for convenient scaling. And, the Silva has a compass capsule that rotates independently of its base. For these reasons, the Silva serves well as a protractor for plotting angles relative to a map's grid.

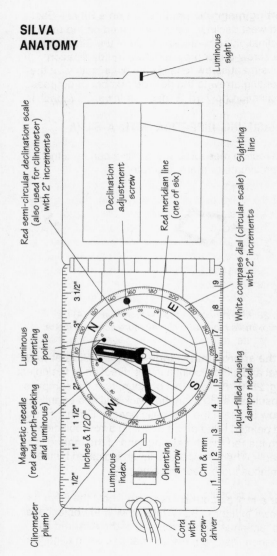

Fig. 27. Silva Ranger model 15 TDCL (graduated 0°–360°)

Setting magnetic declination on a Silva—East and west declinations are indicated on the red declination semi-circular scale of the Silva. Using the screw-driver provided, turn the declination adjustment screw to bring the black outline of the orienting arrow to the desired number of degrees declination (toward 'E. decl.' or toward 'W. decl.').

MEASURING DIRECTION WITH A SILVA

A direction measured with a Silva can be either approximate or more accurate.

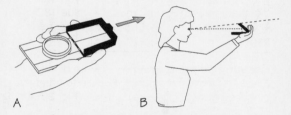

Fig. 28. Measuring direction with a Silva in (A) the approximate mode and (B) the more accurate mode.

In the approximate mode—hold the Silva at waist level and simply point it in the direction of the object (Fig. 28A). While in this position, turn the compass capsule so as to align the black-outlined orienting arrow beneath the magnetic needle. Do this so that the pointed head of the orienting arrow coincides with the red (north-seeking) end of the magnetic needle. Then, read the azimuth indicated by the luminous index near the hinge.

For a more accurate reading—hold the Silva as shown in figure 28B. While viewing the object over the luminous sight at the end of the mirrored lid (dashed line), turn the compass so as to align the sighting line of the mirror (dotted line) with the reflection of either luminous index painted on the compass base. Continue with the procedure described under 'approximate mode.'

MEASURING STRIKE AND DIP—CONTACT METHOD

Measuring strike—Hold the Silva level against an inclined plane (as with the Brunton in figure 19), and measure direction as in the 'approximate mode.' The direction that you measure is strike. (The lack of a round level limits the accuracy of your strike measurement.) Whether you record this direction, or whether you record a value 180° from it, is dictated by the right-hand rule described on page 18.

Measuring dip—In order to measure dip with the Silva, first turn the compass capsule to bring the magnetic declination scale to index (Fig. 29).

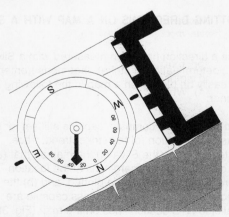

Fig. 29. A Silva showing the semi-circular magnetic declination scale brought to index to allow for measuring the dip of the inclined surface. In this example dip is 30°.

Orientation can be achieved by bringing either E (90°) or W (270°) to either luminous index painted on the base. Inasmuch as the declination scale is fixed to the azimuth scale of the compass capsule, the 90° positions of the declination scale will also be at the luminous indices.

Now hold the Silva against the inclined surface (as with the Brunton in figure 20). The vertical angle (dip) will be indicated by the red clinometer plumb arrow against the magnetic declination scale. (If the semi-circular scale is upside-down, owing to the particular orientation of the Silva, simply turn the compass capsule through 180° to bring it into a readable position. As with a Brunton, you have to experiment with a Silva in this mode until you are satisfied that you have detected the maximum possible vertical angle (i.e., true dip).

PLOTTING DIRECTIONS ON A MAP WITH A SILVA

Once a direction has been measured with a Silva (e.g., determination of strike), that measurement can easily be plotted on a map in the following manner:

Step 1—With the compass capsule still set for the measured direction (of the strike, trend, trace, whatever), place the Silva on the map so that (a) one of its graduated margins is at the location where the direction is to be plotted, and (b) the red meridian lines within the compass capsule are parallel to N–S section lines on the map (Fig. 30).

Step 2—Draw a line representing the direction to be plotted along a graduated margin. That's it!

The graduations along the margins (centimeters/millimeters and inches) can be used to good advantage when plotting a directional leg of a paced or taped traverse. Notice that once a direction has been measured, its plotting does not depend on the position of the magnetic needle, so the map need not be oriented in the field.

As can be seen from the foregoing, the Silva serves as a much more convenient protractor than does the Brunton. Recall from page 11 that for the Brunton to be used as a plotting protractor, first the map must be field-oriented. Not so for the Silva.

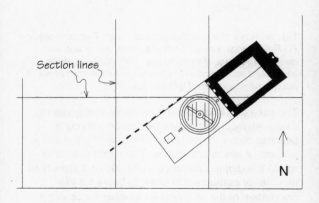

Fig. 30. A Silva positioned on a map for plotting a direction indicated by the dashed line. Notice that the meridian lines within the compass capsule (six in number) are parallel to N–S section lines. Caution! Not all section lines run exactly north–south. You might have to construct a reference line parallel to a side of the map (i.e., parallel to a meridian, which is a true north–south line).

33

USING A SILVA TO MEASURE RAKE (OR PITCH)

The fact that the Silva's compass capsule can be rotated independently of its base provides an easy technique for measuring the rake of a lineation.

Definition—Rake (also called pitch) is the angle that a lineation on a dipping bed makes with a horizontal line in that bed (Fig. 31). (The horizontal line in the bed is, of course, strike of the bed.)

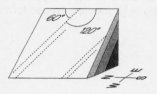

Fig. 31. Rake of a lineation (dotted lines). Two conventions: (1) Rake expressed as an acute angle, along with the quadrant bearing of its trend e.g., (60° NW); (2) Rake expressed as an angle measured clockwise from strike, applying the right-hand rule of page 18 (e.g., 120°).

The rake of a lineation is commonly measured in cases where the lineation occurs on a dipping bedding plane (dotted lines of figure 31), the strike and dip of which are known. The intent is to later rotate the dipping plane (with the aid of a stereographic net or computer) in order to solve for the orientation of the lineation as it occurred before it was structurally tilted.

Step 1—Set the semi-circular magnetic declination scale at index for measuring a vertical angle (as described under 'measuring dip' on page 31), and place the Silva vertically on edge on the inclined surface at or near a lineation (Fig. 32A).

Step 2—With the Silva vertically on edge and resting on the inclined surface, position it so as to bring the plumb arrow to zero on the declination scale. The Silva's edge now marks the strike of the inclined bed on which it is resting (Fig. 32A).

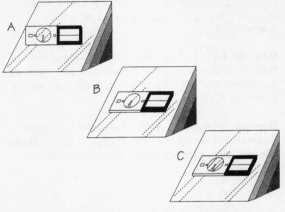

Fig. 32. Measuring the rake of a lineation on a dipping bed with a Silva. Because of the small scale of this figure, only two of the six meridian lines in the compass capsule are shown (and only in C).

Step 3—Using the edge in contact with the dipping bed as a hinge-line, flatten the Silva's base against the bed's surface and hold in place (Fig. 32B).

Step 4—Rotate the compass capsule so as to bring the meridian lines into parallel alignment with the lineation (Fig. 32C).

Step 5—Rake is indicated (on the circular azimuth scale) by one of the two luminous indices painted on the compass base. Which of the two indices should be used depends on the particular convention in practice (Fig. 31).

THE PLANE TABLE AND ALIDADE

There are several varieties of alidades, but figure 34 pretty well covers their general features. The two most common types of Beaman scales are illustrated in figure 33. Their brief explanations at this point are intended to serve as reference material later.

Beaman type 1 (Figs. 33A and 34)—Read the horizontal Beaman to the nearest unit and then, as a percent, multiply it by slope distance (i.e., stadia interval x 100).

Beaman type 2 (Fig. 33B)—Read the horizontal Beaman to the nearest unit, multiply it by the stadia interval, then subtract it from slope distance.

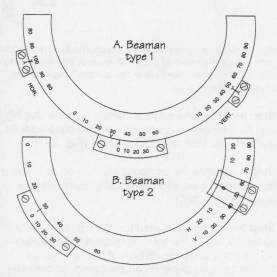

Fig. 33. Two common types of Beaman scales.

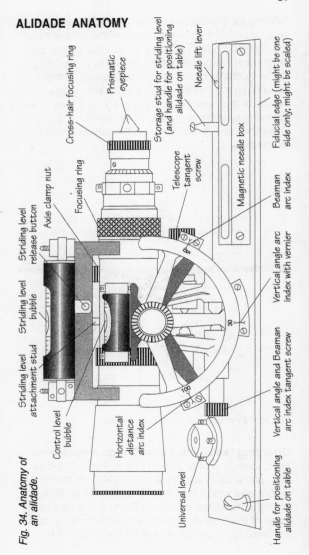

Fig. 34. Anatomy of an alidade.

Setting up the table—Set up the tripod with two legs down-hill if on slope. Tighten the wing-nuts of the Johnson head (Fig. 35), attach the board, and loosen the wing nuts. While grasping the far edge of the board with one hand and bracing its top against your forearm, place the alidade in the center of the board. Level the bull's eye bubble and tighten the upper wing-nut. (The upper wing nut locks the table in a plane, but the table remains free to rotate.)

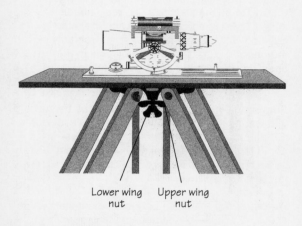

Fig. 35. Wing nuts of the Johnson head.

Orienting table, when working on a base map—First, identify on the base map (a) the location of your plane table station and (b) a feature that is visible and which is at least 4 inches (map distance) from the plane table station. Then, place one long edge of the alidade through these two points, rotate the table to a position where the feature can be sighted through the alidade, and tighten the lower wing-nut. The table is now oriented relative to the surrounding terrain. Incidentally, if your alidade lacks a magnetic needle, you can use a compass to orient the table.

Orienting table, when working without a base map—First, draw a line several inches long representing true north on the plane table sheet. Then, using a protractor, plot and draw an equally-long magnetic declination line coincident with one end of the north-line. Place a long edge of the alidade along the declination line and rotate the table to bring the alidade's compass needle to index. Tighten the lower wing-nut. The table is now oriented with respect to true north. Arbitrarily plot the position of the plane table station so that the sheet will accommodate as much of the contemplated traverse as possible.

Focusing—It is important that the focus be such that when moving your eye slightly (while looking through the instrument) the group of cross-hairs and the rod do not appear to move with respect to each other. Adjust the cross-hair focusing ring as necessary to fix them.

Adjustments—At each new instrument station you should check the adjustment of the vernier level. First, level the instrument with the striding level by turning the right tangent screw. Reverse the striding level and, if the bubble does not return to center, bring the bubble one-half the distance to center with the capstan screws at both ends of the striding level. Next, bring the Beaman index to 50 on the V scale with the left tangent screw. If necessary, the vernier level can then be centered by turning its adjusting screws.

Reading the vertical arc and vernier—The vertical angle arc and arc vernier measure the inclination of the telescope in degrees and minutes (Fig. 36). The angle arc is graduated in degrees and half-degrees, and its index is at zero on the vernier. Minutes of arc are read on the vernier, which is graduated from 0 to 30 minutes. Spacing among graduations on the two scales differs slightly so that only one pair of opposing graduations can be coincident at any one setting. The coincident graduation on the vernier indicates the number of minutes that must be added to the degree or half-degree graduation on the arc scale just to the left of the index. As an example…in figure 36 the arc scale just to the left of the index reads 35 degrees and 30 minutes. The coincident graduation on the vernier is 16 minutes. So…the computed value is 35 degrees and 46 minutes.

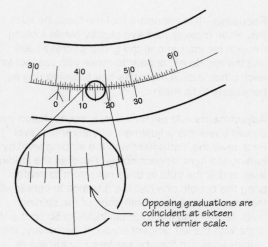

Opposing graduations are coincident at sixteen on the vernier scale.

Fig. 36. Vertical angle scale on upper arc and vernier scale on lower arc.

Stadia distance—The stadia interval (i.e., the rod intercept between the highest and lowest cross-hairs) multiplied by 100 gives the distance in feet between the instrument and the rod. At great distances, or in cases where the rod is partially concealed by terrain, it may be necessary to read one-half (or one-quarter) stadia interval and multiply by two (or by four) to derive a full stadia interval (Fig. 37).

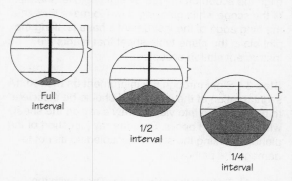

Fig. 37. Applying different stadia intervals as a function of distance.

For maximum accuracy, place one cross-hair at an even footage increment and count toward the fractional increment (Fig. 38). The stadia distance is the slope distance between the instrument and the rod, but it's the horizontal distance (i.e., map distance) that must be plotted. Of course, in cases of horizontal shots the two are equal.

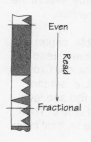

Fig. 38. Reading the rod from even footage to a fractional increment. Interval: 1.65 feet.

Procedure—'running the gun.'

Step 1—When beginning at an instrument station whose elevation is known, record its number and its elevation under Instrument Station, and record the height of the instrument (H.I.) under Remarks.

Step 2—With the instrument axis clamp loosened so that the telescope barrel can be manipulated, align the scope on the rod by sighting over the top of the scope while glancing down to make sure that the long edge of the instrument's base is along the pinhole in the plane table sheet that marks the instrument station occupied.

Step 3—Sight through the instrument on the rod. (A slight rotation of the instrument should be sufficient to bring the rod into view.) Draw a ray on the sheet with a 8H or 9H pencil. Stop the ray just short of the pinhole marking the station occupied so as not to damage the pinhole.

Step 4—Tighten the axis clamp. Then, with the right tangent screw, bring a horizontal cross-hair to an even footage increment on the rod. Count the stadia interval to the nearest (estimated) one-hundredth of a foot and record under Stadia Interval. If converting to a full stadia interval (from a 1/2 or a 1/4 stadia interval) is necessary, record the whole interval after making the conversion.

Step 5—If a horizontal shot is possible, center the striding level with the right tangent screw, read the middle hair on the rod, and record it under Cross Hair. If a horizontal shot is not possible, set the middle hair on the mid-region of the rod and level the vernier level with the left tangent screw. Then, using the right tangent screw, set the vertical index on the nearest graduation and record that graduation under V. Read H and record under H. Read the middle hair on the rod to the nearest one-tenth of a foot and record under Cross Hair. (Assign a negative value to the cross-hair reading when determining the elevation of a rod point, and a positive value when determining the elevation of a new instrument station.)

Step 6—Wave off the rod man so that he/she can immediately pursue the next rod station.

Step 7—Multiply H x stadia distance (stadia interval x 100), record under Horizontal Distance, and scale this distance along the ray. Mark the rod station with a pin prick and label it with its number.

Step 8—Subtract 50 from V reading, record under V factor, and multiply by stadia interval and record under V factor x Stadia Interval. Maintain sign of the latter value when determining elevation of the new rod point. When determining elevation of a new instrument station, change the sign and encircle the sign to designate the change.

Step 9—Algebraically add the cross-hair reading (negative when determining elevation of a rod point, R.P., and positive when determining elevation of an instrument station, I.S.) and the instrument height (positive when determining elevation of a rod point, and negative when determining elevation of an instrument station) to the vertical difference between the instrument and the point on the rod (V factor x stadia interval). Algebraically add this value to the elevation of the earlier point and record under New Elevation Derived (R.P. or I.S.).

Computing elevations—foresights—Solving for the elevation of a new rod point (Elev. R.P.) from an established instrument elevation (Elev. I.) can be either positive, i.e., shooting uphill (Fig. 39A), or negative, i.e., shooting downhill (Fig. 39B).

A. Positive foresight

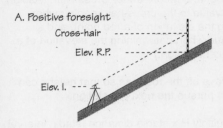

Elev. R.P. = Elev. I. + [(V − 50) × Stad. Int.] − Cross-hair

B. Negative foresight

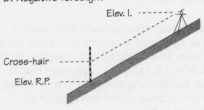

Elev. R.P. = Elev. I. + [(V − 50) × Stad. Int.] − Cross-hair

FIG. 39. Graphic portrayal of positive and negative foresights.

Computing elevations—backsights—Solving for the elevation of a new instrument station (Elev. I.) from an established rod point elevation (Elev. R.P.) can be either positive, i.e., shooting uphill (Fig. 40A), or negative, i.e., shooting downhill (Fig. 40B).

A. Positive backsight

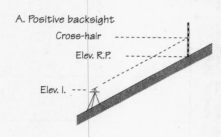

Elev. I. = Elev. R.P. − [(V − 50) × Stad. Int.] + Cross-hair

B. Negative backsight

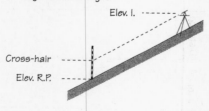

Elev. I. = Elev. R.P. − [(V − 50) × Stad. Int.] + Cross-hair

Fig. 40. Graphic portrayal of positive and negative backsights.

Fig. 41. Format for recording observations and computations.

Inst. Sta.		Rod Point		Stadia x 100 =		Beaman Scale		Cross Hair	Horiz. Dist.	V factor (V – 50)	V factor x Stadia Interval	New Elev. Computed (R.P. or I.S.)	Remarks (Inst. Ht. where needed, etc.)
No.	Inst. Elev.	No.	R.P. Elev.	Interval	Distance	V	H						

THE STEREOGRAPHIC NET

Purpose—The stereographic-net method is used to represent three-dimensional orientations of planes and lines in two-dimension, and to illustrate relationships among such objects. This is done by placing the plane or line centrally within a sphere so that it passes through the sphere's center. Unlike mineralogists, who use the upper hemisphere for projection purposes, structural geologists use the lower hemisphere. For example, let's represent the inclined plane of figure 42A, which strikes due N and dips 50° W. Inasmuch as the plane passes through the center of the sphere, its intersection with the sphere forms a 'great circle' (i.e., a circle that bisects a sphere). That part of the great circle in the lower hemisphere (the part bounding the shaded part of the inclined plane in figure 42A) is projected to the equatorial plane (marked with N–S and E–W axes) by constructing lines between the great circle and a zenithal point (north pole of the sphere), as in figure 42B. This projection is shown in two-dimension as the 'cyclogram' of figure 42C. Viewing figure 42C is like looking into a bowl that has the cyclogram drawn on its interior surface.

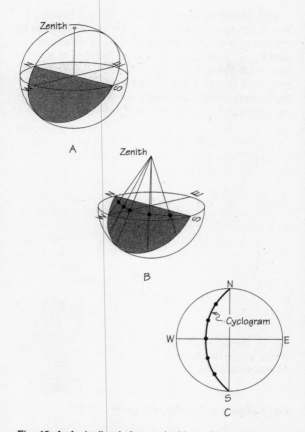

Fig. 42. A: An inclined plane coincident with a sphere and passing through its center. Its shaded one-half is within the lower hemisphere. B: Lines running from the margin of the sphere-bounded plane, through the equatorial plane (defined by the N–S and E–W axes) to the sphere's zenith. Dots marking the intersections of the lines with the equatorial plane describe a cyclogram. C: The equatorial plane (in the plane of the page) with the cyclogram defining the orientation of the plane.

Construction—A stereographic net is constructed by projecting a family of 'great circles' (analogous to meridians or lines of longitude) and 'small circles' (analogous to parallels or lines of latitude) to an equatorial plane (Figs. 43A and 43B). Great circles and small circles in figure 43B are graduated in increments of 30°. A stereographic net graduated in increments of 10° is shown in figure 43C.

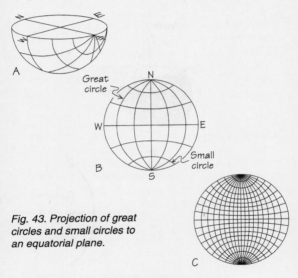

Fig. 43. Projection of great circles and small circles to an equatorial plane.

PREPARING A STEREOGRAPHIC NET

A standard stereographic net diagram can be prepared for use by pushing a thumb-tack from below through its center. (The tack can be secured to the bottom with tape.) A transparent overlay is pushed onto the tack so that it can be rotated freely. The overlay should be labeled N, S, E, and W at points coincident with those of the net. The overlay is said to now be in its 'coincident position.'

PLANAR AND LINEAR FEATURES

Planar features include both primary features (e.g., cross-beds) and secondary features (e.g., foliations). Similarly, linear features include both primary features (e.g., sole-marks) and secondary features (e.g., hinge lines of folds).

REPRESENTING AN INCLINED (DIPPING) PLANE

Example: a plane striking N 30 E and dipping 50 NW (Fig. 44)—Note: on diagrams, such as figure 44, the dotted N–S and E–W axes represent the fixed orientation of the underlying stereographic net; the N,S,E, and W labels are on the overlay and indicate the degree to which the overlay is rotated.

Step 1 (Fig. 44A)—With the overlay in its coincident position, using small circles (a) count off 30° from north to east along the perimeter of the net. Mark this position, and (b) extend a line from it through the net's center to the opposite side of the perimeter. This line, with the overlay in coincident position, represents the strike of the plane.

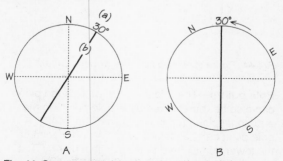

Fig. 44. Stereographic representation of an inclined (dipping) plane with a cyclogram. (Figure continued on the following page.)

Step 2 (Fig. 44B)—Rotate the overlay counter-clockwise so as to bring the strike line into coincidence with the net's N–S axis.

Step 3 (Fig. 44C)—Inasmuch as the plane dips NW, using the longitudinal circles, count off 50° from west to east along the W–E axis. Trace the longitudinal circle that occurs at this 50° point onto the overlay, making a cyclogram. (Ignore pp for the moment.)

Step 4 (Fig. 44D)—Rotate the overlay clockwise back to its coincident position. The cyclogram representing the inclined plane is now in its definitive position.

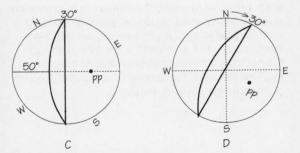

Fig. 44 (Figure continued from the previous page).

Pole points—The orientation of an inclined plane can also be represented by a 'pole point' on the equatorial plane (pp in figures 44C and 44D). A pole point can be plotted by projecting a point on the lower hemisphere to the zenithal point. (That point on the lower hemisphere is where the hemisphere is struck by a line that is perpendicular to the inclined plane at the center of the net.)

In order to plot the pole point of the inclined plane illustrated in figure 44, repeat steps 1 and 2, but then, instead of counting 50° inward along the W–E axis from the western perimeter of the net (as you did in step 3), count 50° from west to east along the W–E axis beginning at the center of the net. Upon rotating the overlay back to its coincident position, the pole point is in its definitive position. The pole point for this inclined plane (with strike N 30 E and dip 50 NW) is shown both in figure 44C and in figure 44D.

REPRESENTING AN INCLINED (PLUNGING) LINE

Example: a line plunging ('dipping') S 20 E at a vertical angle of 60 (Fig. 45).

Step 1 (Fig. 45A)—With the overlay in its coincident position, using the small circles, count off 20° from south to east along the perimeter of the net. Mark this position.

Step 2 (Fig. 45B)—Rotate the mark to the nearest axis, in this case clockwise to the net's S pole.

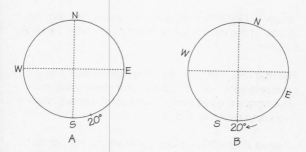

Fig. 45. Stereographic representation of an inclined (plunging) line. (Figure continued on the following page.)

Step 3 (Fig. 45C)—Count 60° from the net's S pole up the N–S axis and mark this point.

Step 4 (Fig. 45D)—Rotate the overlay back to its coincident position.

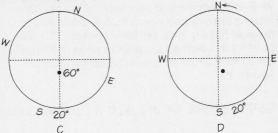

Fig. 45 (Figure continued from the previous page).

REPRESENTING THE PLUNGE OF A LINE FROM ITS RAKE IN AN INCLINED PLANE

Example: a plane striking N 45 E and dipping 60 SE, on which there occurs a line with a rake of 40 in a general SW direction (Fig. 46 block diagram). The cyclogram for the inclined plane has already been constructed.

Step 1 (Fig. 46A)—Rotate the overlay to bring the strike of the cyclogram to coincidence with the N–S axis.

Step 2 (Fig. 46B)—From the south pole of the stereographic net, count 40° (the rake) along the cyclogram's great circle, using small-circle graduations to measure the 40°.

Step 3 (Fig. 46C)—Extend a line from the center of the stereo net through the 40° point to the perimeter of the stereo net.

Step 4 (Fig. 46D—Rotate the overlay so as to restore coincidence between the overlay and the stereo net. The exact bearing of the line can be read at the point where the constructed line intersects the net's perimeter, using small-circle increments to measure. Bearing: S 25 W.

Step 5 (Fig. 46E)—If the line constructed in step 3 is rotated to a principal diameter of the stereo net (i.e., either the N–S axis or the E–W axis), the plunge of the line can be read, measuring from the perimeter inward toward the center.

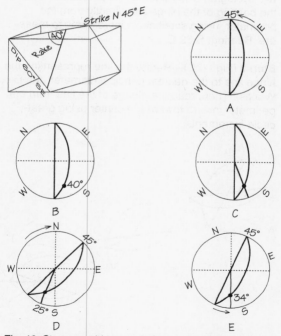

Fig. 46. Stereographic net solution of the rake (pitch) of a line within an inclined plane.

REPRESENTING A LINE FORMED BY THE INTERSECTION OF TWO PLANES

Example: Two cyclograms representing the opposing limbs of a fold have already been constructed (Fig. 47A). Problem: solve for the bearing and plunge of the fold's 'hinge line' (i.e., the line of intersection of the fold's two limbs).

Step 1 (Fig. 47B)—Draw a line from the center of the stereographic net, through the point where the two cyclograms intersect, to the perimeter. Read the bearing of the hinge line directly from the perimeter, using small-circle increments to measure. Bearing: S 62 E.

Step 2 (Fig. 47C)—Rotate the line representing the hinge line to the nearest principal diameter (the E–W axis). Then, read the plunge of the line from the perimeter inward toward the center using great-circle increments.

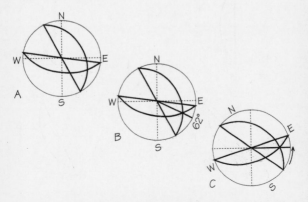

Fig. 47. Stereographic net solution for the orientation of the line formed by the intersection of two planes.

SOLVING FOR ORIGINAL ORIENTATION OF STRUCTURALLY TILTED CROSS-BEDS

This procedure consists of first plotting both the orientation of major stratification and that of an included cross-bed on a single overlay. Then, the major stratification is rotated about a horizontal axis, producing the cross-bed's original orientation. Orientations of both the major stratification and the cross-bed are represented simply by azimuths and magnitudes of their dips. (Review the azimuth method explained on page 19.)

Example: The major stratification dips with an azimuth of 40 and a magnitude of 30. An included cross-bed dips with an azimuth of 70 and a magnitude of 60.

Step 1 (Fig. 48A)—Plot and label a pole point for major stratification, s, and for the included cross-bed, c. Review pole points, bottom of page 52, top of page 53 (including figures 43-C, D) if needed.

Step 2 (Fig. 48B)—Rotate the overlay to bring the pole point for major stratification, s, to E–W axis.

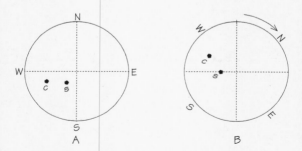

Fig. 48. Solution of original orientation of a structurally tilted cross-bed. (Figure continued on the following page.)

Step 3 (Fig. 48C)—Rotate the major stratification to its original horizontal position by migrating its pole point along the E–W axis to the center of the net, in this case 30°.

Step 4 (Fig. 48D)—Without moving the overlay, migrate the pole point of the cross-bed along a small circle by the same amount as that required to bring s to the center of the net (i.e., 30°). Label the migrated cross-bed pole point c'.

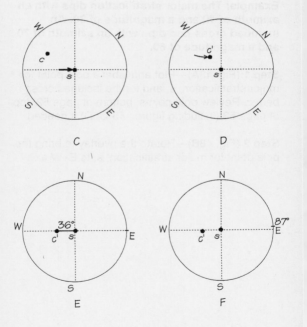

Fig. 48 (Figure continued from the previous page).

Step 5 (Fig. 48E)—Rotate c' to the E–W axis and place a tick-mark at the E pole. Read the original dip magnitude of the cross-bed along the E–W axis, beginning at the net's center and ending at c' (i.e., 36°).

Step 6 (Fig. 48F)—Rotate the net to coincidence and read the original azimuth of the cross-bed's dip direction at the tick-mark (i.e., 87°).

Note—In some cases the cross-bed's pole point migrates off of the overlay when the major stratification is rotated. When this happens, follow the procedure described on the following page and which is illustrated in figure 49.

Incidentally, the procedure described here for the rotation of cross-beds can also be used to solve for the pre-unconformity orientation of layers that were (a) deposited, (b) inclined, (c) partially eroded, (d) covered by younger layers, and (e) inclined further.

60

If rotation of the major stratification about a horizontal axis causes the cross-bed's pole point to migrate off of the overlay (Fig. 49), continue rotating the pole point in the same direction, and along the same small-circle value, but at a place 180° along the perimeter of the stereographic net.

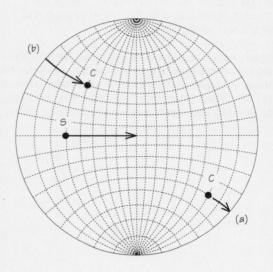

Fig. 49. The major stratification pole point, s, has been migrated 60°, causing the cross-bed's pole point, c, to migrate off the diagram at (a) after 20° migration. Solution: migrate c an additional 40° along the same small-circle — but 180° removed (b)—to complete a migration equal to that of the major stratification (i.e., 20° + 40° = 60°).

SOLVING FOR STRIKE AND TRUE DIP WITH TWO APPARENT DIPS

Inasmuch as a line of apparent dip lies within the dipping plane, two such lines can define the orientation of that plane (Fig. 50).

First, plot a point representing the bearing and plunge of a line of apparent dip on the overlay—as already explained on pages 53 and 54; example: a line of apparent dip plunging N 50 E at 20. Next, plot a second line of apparent dip plunging S 30 E at 35.

Step 1 (Fig. 50A)—Rotate the two points to a position where they both fall on the same great circle. Trace that great circle onto the overlay, creating a cyclogram. Read the true dip from the E perimeter inward toward the cyclogram: 42°.

Step 2 (Fig. 50B)—Return the overlay to its coincident position and read the strike: N 25 E. The direction of true dip can now be seen by inspection to be SE.

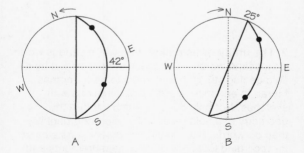

Fig. 50. Solving for strike and dip with two apparent dips.

CONFRONTING AN OUTCROP—points to ponder

1. The value of reconnaissance—Before taking pencil and notebook in hand and putting your brain in low-gear, invest some time in cruising around your study area. Is there a field geologist who hasn't spent valuable time at a particular location (describing, sketching, interpreting, etc.), only to later learn that just around the bend there is an exposure more enlightening and illustrative? (Curses!)

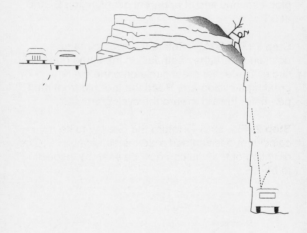

2. The utility of field sketches—'A picture is worth a thousand words.' (No doubt first exclaimed by a field geologist!) You don't have to be a Thomas Moran to enhance your field descriptions with sketches. Not only does a sketch capture the geometry of a field relationship more thoroughly than do words, but a sketch also serves as a base for recording locations of field measurements and rock samples. Lastly, and universally, sketching requires that you observe the object more thoroughly, thereby prompting discovery. In this last way, sketching is superior to using an instant camera.

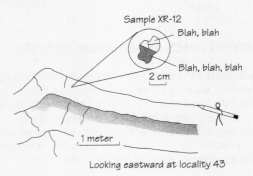

3. The importance of contact relationships—The end purpose of most field studies is to decipher the geologic history of an area, and geologic history is punctuated by contact relationships. Observe, sketch, and record the geometry of contacts and the character of rocks on either side. The variety of possible features is too numerous to list here, but examples include:

In sedimentary rocks—Contacts: abrupt, gradual, truncating (both macro- and micro-truncation). Underlying intervals: weathered, mineralized, penetrated. Overlying intervals: sole-marked, graded, conglomeratic.

In igneous rocks—Alteration effects (e.g., chilled, baked, micro-intruded). Inclusions. Injections.

MAPPING—PACE-AND-BRUNTON METHOD

Definition—Pace-and-Brunton mapping is a 'quick-and-dirty' method of mapping features within a relatively small area (i.e., a size comparable to that of a football field or less). The map is made by 'stepping-off' a series of distances, each of which is in a direction (either bearing or azimuth) measured with a compass, such as the Brunton.

Procedure—First, you will need to know the length of your pace, which can be determined by counting the number of paces required to cover a known distance (e.g., a distance of 100 feet staked out with a tape). Pace is conventionally considered to be two steps, so count every other step (i.e., every right step or every left step) along the traverse.

Example—A pace-and-Brunton map is simply a plot of the positions of selected points on the ground, either natural or man-made, constructed from bearings or azimuths and paced-distances from point to point (Fig. 51). Data (employing the azimuth method) plotted in figure 51 are as follow:

Leg	Azimuth	Distance
A–B	61°	145'
B–C	94°	95'
C–D	160°	80'
D–E	139°	50'
E–F	262°	245'
F–G	337°	90'

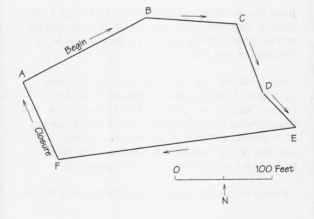

Fig. 51. A plot of the pace-and-Brunton data shown above.

Correcting error of closure—A traverse constructed by the pace-and-Brunton method—or even that of the tape-and-Brunton method—will probably result in some amount of error of closure. That is, an attempt to return to your original starting point will usually fail to precisely close the loop (Fig. 52A). Errors occur in determining both distance and direction among points; and, they are cumulative. The following procedure is an acceptable method for correcting an error of closure.

Step 1—Divide the distance between A' and A into n equal segments, with n = number of bi-directional points called 'turning points.' Exclude A and A'. In the case of figure 52, $n = 3$.

Step 2—Move every turning point in a direction parallel to A'–A (in order to close on A) by an amount equal to a fraction of the distance A'–A. The denominator of the fraction consists of the number of segments within A'–A, and the numerator consists of the number of the particular turning point. The fractions are shown in figure 52B.

Step 3—Connect all adjusted turning points and A.

67

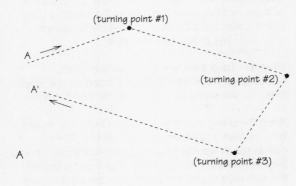

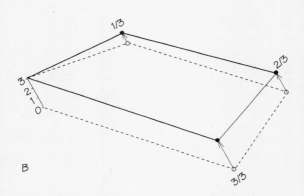

Fig. 52. (A) Preliminary drawing (dashed line). Compass directions and numbers of paces were chosen so that A' should have closed on A; but, it didn't. (B) Final drawing (solid line) after correcting the error of closure.

TRIANGULATING

Intersection method—In the field it is possible to plot the position of an unoccupied point on a map using a triangulation method called intersection (Fig. 53). The procedure is simple:

Step 1—Take the bearing or azimuth of the unoccupied point from Point X plotted on the map. Then do the same for Point Y plotted on the map.

Step 2—Using a protractor aligned with the map's grid, project a ray from each of the two plotted points (each with its bearing or azimuth) toward and beyond the unoccupied point. The intersection of the two rays marks the position of the unoccupied point. Note: For maximum accuracy, the angle formed by the two intersecting rays (angle 'a' in figure 53) should fall between 60° and 90°.

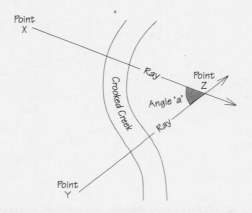

Fig. 53. Solving for the position of an unoccupied point (Z) using the intersection of rays projected from two points (X and Y) already plotted on the map.

Resection method—The resection method of triangulation is the reverse of the intersection method. It solves for the position of your occupied point by triangulating on two other observable points that are shown on your map (e.g., houses, mine workings, bridges, etc.). Simply take a bearing or azimuth on each of the observed points and, using a protractor aligned with the map's grid, draw a ray on the map extending from each of the points toward you. The intersection of the two rays marks your position. Note: Each bearing or azimuth that your project must be 180° from the value read with the white (north-seeking) end of the needle, because you are projecting it backward. Or, more easily, read the bearing (or azimuth) of the black end of the needle. Once again, 60°–90° between the two intersecting rays is best for accuracy.

In some instances, when your field position is along a linear feature shown on the map (e.g., a road or river), you can determine your position by projecting a single ray (Fig. 54).

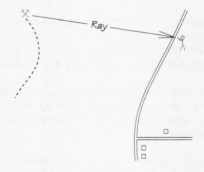

Fig. 54. Plotting your field position by projecting a single ray. You know that you are along a road, but you don't just where. Sighting on a mine workings shown on your map can provide your precise location.

DESCRIBING STRATIGRAPHIC SECTIONS—FORMAT

Lithic descriptions:

There are as many schemes for lithic descriptions as there are geologists. A few guiding principles:

A. List the most important lithic features first, followed by less important ones. Try to organize your format the way in which a person might think about the physical character of a rock unit.

B. Use punctuation (caps, dashes, semi-colons) to emphasize your lithic hierarchy.

C. Be consistent. Come to think of it, this principle is not a bad one to apply to everything you do.

Some example-formats follow:

Sedimentary rocks—Principal rock name, compositional adjectives; color, grain size(s), degree of induration; minor constituents; fossils. Bedding thickness, notable features. Nature of boundaries.

Igneous rocks—Principal rock name, compositional adjectives; color, crystal size(s); minor constituents. Notable fabric features. Field occurrence. Nature of boundaries.

Metamorphic rocks—Principal rock name, compositional adjectives; color, crystal size(s); accessory minerals and their textures. Details of any foliations and lineations. Field occurrence. Nature of boundaries.

Geomorphic profile:

In cases of sedimentary rocks especially, it is highly communicative to include a geomorphic profile (i.e., a weathering profile) when illustrating a stratigraphic section (Fig. 55). Carefully-drawn geomorphic profiles can greatly facilitate correlating among sections. Tip: The closer that the scale of the field drawing is to that of the final illustration, the easier will be the transcribing.

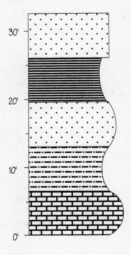

Fig. 55. A schematic example of a geomorphic profile. Lithic-symbol information is enhanced by a weathering profile.

CONSTRUCTING A GEOLOGIC ROAD TRAVERSE

The object of this exercise is to construct a reconnaissance map of outcrops along a road (Fig. 56). This example employs the pace-and-Brunton technique introduced on pages 64–65, only now you are asked to add structural measurements and lithic descriptions.

Each paced leg is along a line-of-sight, the length of which is limited by the curvature of the road. A sample format follows below, from which the map on the facing page has been plotted. The length of pace for this geologist is 6 feet. The method of recording strike and dip is according to the right-hand rule explained top of page 18.

Location of traverse: _____ Date: _____

Leg	Az.	#paces	Item	Description
A–B	306	0	A	Beginning pt.; congl.
		100	Strat. contact	Congl. underlain by shale; strike 70, dip 30.
		166	B	- - - - - - - - - - - -
B–C	7	175	Strat. contract	Shale underlain by sandst.; strike 70, dip 33.
		250	Strat. contact	Sandst. overlain by shale; strike 250, dip 45.
		255	C	- - - - - - - - - - - -
C–D	30	300	Strat. contact	Shale overlain by congl.; strike 250, dip 43.
		350	D	- - - - - - - - - - - -

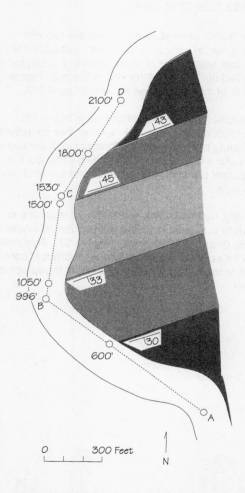

Fig. 56. Reconnaissance map plotted from data on the facing page. Note: Paces in the table on the facing page have been converted to feet on this map.

USING CONTOUR MAPS

The 'rules of contours'—There are several fundamental principles (or 'rules') associated with contour lines. Eight of these rules are illustrated and spelled out in figure 57B, which is a map representation of the perspective shown in figure 57A.

In addition to the rules illustrated in figure 57, contours do not cross one another, nor do individual contours branch. Apparent branching of contours can occur in cases of overhanging cliffs, but the apparent branches are separate contours, differing in value.

Finally, by convention, every fifth contour line is usually a bit heavier in line weight than are others (See 0- and 100-foot lines in figure 57). This makes it a bit easier to visually follow an individual contour, and lends a more obvious fabric to a topographic map.

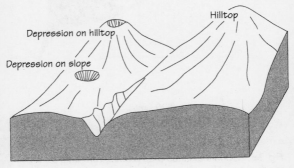

A

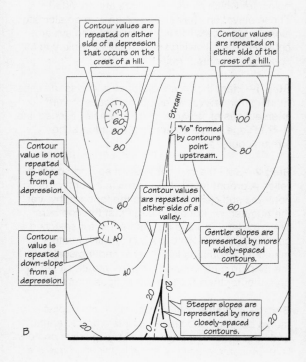

Fig. 57. The 'rules of contours.' The figure on this page is a map representation of the perspective shown in A on the facing page.

CONSTRUCTING A CONTOUR MAP

Figure 58 (facing page) shows the distribution of major streams (dot-dash lines) and a group of points for which elevation values are provided. These elevations have been taken with an altimeter along ridge crests (drainage divides) and stream courses. The following steps should enable you to construct a contour map:

Step 1—Mark the crests of ridges with dashed lines running from elevation points on ridges toward intersections of streams. (See example, dashed line A–B.) Ridge crests bisect uplands separating stream courses.

Step 2—Select a contour interval appropriate to the relief. A proper interval will result in a number of contours sufficient to illustrate the topography, yet the number will not be so large that the contours are crowded. There is no simple rule for selecting a contour interval. Contour spacing will be determined by a combination of relief and the scale of the map. In a hand-drawn map, contours closer to one another than 1/8 inch are tedious to draw.

Step 3—With the contour interval in mind, place marks between successive points along streams and ridges marking positions of prospective contour lines. Assume a uniform slope between points (i.e., equal spacing of contours).

Step 4—Connect marks of equal value with contour lines. (See partially completed solid line C–D.) In order to make your map realistic, curve the contours in a U-shape where they cross noses of ridges, and shape them into abrupt Vs where they cross streams. (See figure 57.)

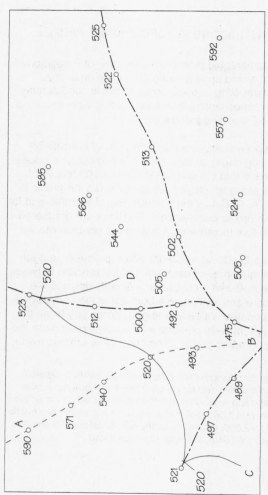

Fig. 58. Exercise in constructing a contour map.

CONSTRUCTING A TOPOGRAPHIC PROFILE

A topographic profile is a side view of topography as it would appear along some line on a map. Constructing a topographic profile is considerably more mechanical than constructing a contour map. The procedure follows:

Step 1—Complete the contour map in figure 59 (facing page) by supplying missing contour values. (Notice that the contour interval is 100 feet.) The line along which you are to construct the profile is line A–B. Below the contour map is a profile-grid for your profile construction. Grid lines are graduated in 100-foot increments, equal to the contour interval.

Step 2—Using a straight-edge (or, better, a right-triangle), project each point of intersection between line A–B and a contour line directly down to the profile-grid and place a bold dot at the corresponding contour value on the grid. When you finish the length of A–B, connect all the dots to complete the topographic profile. (The first 5 dots are supplied.)

Fig. 59 (facing page). A contour map (partially labeled) and a profile grid for constructing a topographic profile along line A–B. Note that the vertical scale on the grid is 1 inch to 1,000 feet, which equals a representative fraction of 1:12,000. Inasmuch as the horizontal scale is smaller (i.e., 1:62,000), the relief is exaggerated.

Contour map

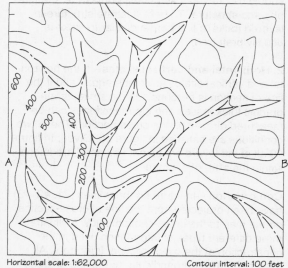

Horizontal scale: 1:62,000   Contour interval: 100 feet

Profile grid

CONSTRUCTING A GEOLOGIC CROSS-SECTION

1. Direction of line of section—A cross-section should be drawn so as to best portray the stratigraphy and structure. To best illustrate structure in layered rocks, a line perpendicular to strike is usually preferred.

2. Horizontal and vertical scales—If the horizontal and vertical scales are made equal, and if the line of section is perpendicular to strike, stratigraphic thicknesses and dip magnitudes can be measured directly on the cross-section. These two conditions are satisfied in figure 60. Given: In figure 60 all of the layers are plane; i.e., none are folded within the area of the map.

Construction—In figure 60 three lines (#1, #2, #3) have been projected from the line of cross-section on the map to the cross-section below to illustrate three different types of information used in the construction of the cross-section. Important: All lines are drawn perpendicular to the line of cross-section (Z–Z') on the map.

Line #1—is projected from where the 400' contour line intersects the line of cross-section, to the 400' value on the vertical scale of the cross-section below. This is one of three such points that can serve as control points for drawing the topographic profile. (Others are at 300' and 500'.)

Line #2—is projected from where a lithic boundary (i.e., between Y and B) intersects the line of cross-section, to the topographic profile. This point on the topographic profile is a control point for drawing the geologic cross-section.

81

Line #3—is extended from a 'strike line' (i.e., the line connecting the two points where the lithic boundary between Y and B intersects the same 300' contour), to that contour's value on the vertical scale of the cross-section below. This point on the cross-section provides the location of the lithic boundary below the topographic profile and illustrates the dip of that lithic boundary.

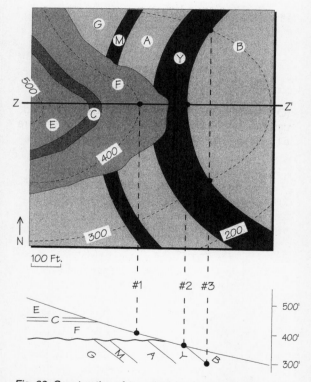

Fig. 60. Construction of a geologic cross-section along line Z–Z'. Given: Layers are not folded within the map area. Vertical and horizontal scales are equal.

SOLVING FOR EXAGGERATED DIP WHEN THE VERTICAL SCALE IS EXAGGERATED

Vertical and horizontal scales of a geologic cross-section should be make equal. If not, dips and stratigraphic thicknesses will be exaggerated. If an exaggerated vertical scale is absolutely necessary, the resultant exaggerated dip can be computed:

tan exaggerated dip (example in figure 61)
= (vertical exaggeration) (tan true dip)
= (2) (tan 15°)
= (2) (.26795)
= .53590
Exaggerated dip = 28°11'

(Of course the easiest way of plotting exaggerated dips and thicknesses is the way I constructed figure 61...s t r e t c h the vertical axis with a computer.)

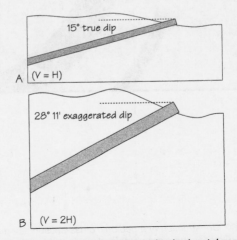

Figure 61. (A) Vertical scale is equal to horizontal scale. (B) Vertical scale is twice horizontal scale.

SOLVING FOR TRUE DIP WHEN A CROSS-SECTION IS NOT PERPENDICULAR TO STRIKE

Trigonometrically—Where a line of cross-section diverges from strike at some angle (0°–90°), true dip can be computed trigonometrically:

tan apparent dip =
    (tan true dip) (sin divergent angle from strike)

Graphically—Figure 62 illustrates W.S. Tangier Smith's graphic solution of true dip when apparent dip and the divergent angle (from strike) of the cross-section are known.

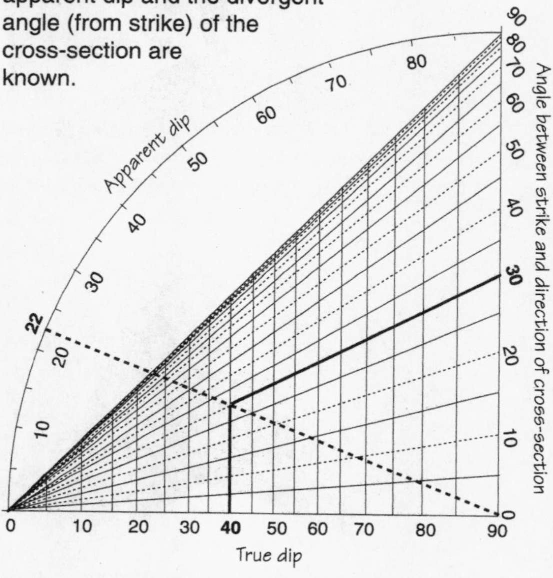

Fig. 62. Graphic solution of true dip. In this example apparent dip is 22°. A line from this value is drawn to the apex of the diagram. The cross-section is 30° from strike. The intersection of the 30°-line and the line to the apex is projected downward to a true dip of 40°. (From Economic Geology, 1925, v. 20, p. 182, fig. 28).

LAND SURVEY SYSTEM

Latitude and longitude—Figure 63 shows the global distribution of latitude and longitude.

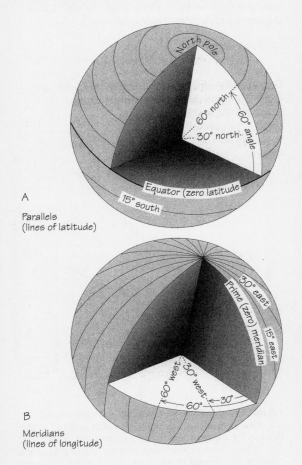

Fig. 63. Global distribution of latitude and longitude.

Townships and ranges—Township and range designations have as their bases a combination of principal meridians and arbitrary base lines (Fig. 64).

Subdivisions of a township are shown in figure 65.

Fig. 64. An example, from the Midwest, of how a meridian (the 5th principal meridian) serves as a basis for townships and ranges. Townships in the shaded area are numbered with reference to a base line in Arkansas and the 5th principal meridian in Arkansas and Missouri.

86

Subdivisions of a township—Figure 65 shows subdivisions of township T1N, R4E.

Sections an parcels—A township consists of a square block of 36 sections, each of which is one square mile (= 640 acres) (Fig. 65). Sections are delineated by section lines (solid where surveyed and dashed where approximate), and each bears its number in its center. Numbering of sections within a township employs a rather curious zig-zag system, beginning in the upper right corner with 1 and ending with 36 in the lower right corner. When describing the location of a parcel of land within a section, the section can be sub-divided into quarters—and further into halves of quarters—and further into quarters of quarters (Fig. 65).

Notice that the order of fractions of a section is as though the word 'of' were inserted between fractions in place of commas. A point, rather than a parcel, can be described as the center of some fraction of a section. Example: point x in figure 65.

In 1979 the U.S. Bureau of Land Management adopted a new way of writing legal descriptions of land. This new method places township, range, and principal meridian before the parcel. For example, parcel B in figure 65 would now be designated by BLM as...

> T. 1N., R. 4E., 5th P.M.,
> Sec. 32, SW 1/4 NE 1/4.
> (Note: no comma between the quarters)

As an alternative to describing a point as being at the center of some fraction of a section, its location can be designated by footage from the north line, and footage from a sideline, of a parcel.

87

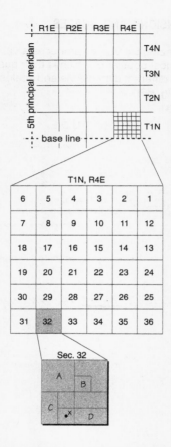

Fig. 65. Each of he parcels of land (A through D in section 32) can be specified by (a) township, (b) section, (c) fraction of section and (d) reference principal meridian (P.M.) as follows:

A—NW 1/4, Sec. 32, T1N. - R4E., 5th P.M.
B—SW 1/4, NE 1/4, Sec. 32, T1N. - R4E., 5th P.M.
C—W 1/2, SW 1/4, Sec. 32, T1N. - R4E., 5th P.M.
D—S 1/2, SE 1/4, Sec. 32, T1N. - R4E., 5th P.M

MAP DIMENSIONS AND SCALES

The U.S. Geological Survey produces a variety of maps, the most popular of which is their quadrangle series. Five different dimensions and scales are available (Fig. 66).

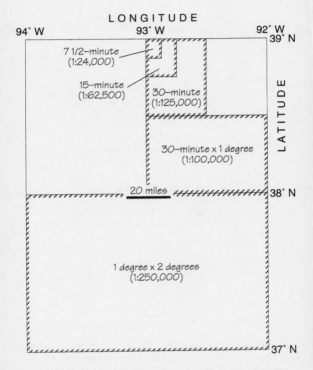

Fig. 66. Dimensions and scales of U.S. Geological Survey quadrangle maps. Latitude and longitude dimensions are given (in minutes or degrees in multiples of two), along with fractional scales. The reason that the fractional scales are not also multiples of two is because the sheets of paper on which these maps are printed differ in size.

Map dimensions—The name 'quadrangle' stems from the fact that the earliest sizes (7 1/2-minute, 15-minute, and 30-minute) are 'square' with respect to latitude and longitude. That is, each is bounded on all four sides by an equal number of minutes. The reason that these three map-sizes appear rectangular in figure 66 is because meridians, which define longitude, converge toward the poles.

Map scales—The scale of a map shows the relationship between the length of a line on a map and the length of that same line on Earth's surface. This relationship can be conveyed by a fractional scale, which is expressed either as a ration or a representative fraction (RF), setting the length of the line on the map equal to one. For example, a scale of 1:24,000 (or 1/24,000 RF) means that one unit on the map (one inch, on foot, etc.) represents 24,000 units on the ground (24,000 inches, 24,000 feet, etc.). A 3-inch line on a map with a fractional scale of 1:24,000 represents a distance on the ground of...

$$3 \times 24,000 = 72,000 \text{ inches}$$
$$\text{divided by 12 (inches per foot)} = 6,000 \text{ feet}$$

A relatively large fraction, for example 1:24,000, is called a 'large-scale' map; and a relatively small fraction, for example 1:250,000, is called a 'small-scale' map. A large-scale map represents a smaller land area in greater detail than does a small-scale map on the same size paper.

Another kind of scale is the graphic scale, which has two advantages over the fractional scale: (1) A graphic scale can be applied visually, rather than arithmetically; and (2) a graphic scale can be enlarged or reduced along with the map and still be applicable. (If you enlarge or reduce a map, its fractional scale no longer applies.) Figure 67 illustrates a two-step procedure for determining distance between two points on a map with a graphic scale.

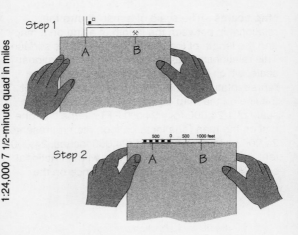

Fig. 67. Measuring map distance with a graphic scale. Step 1: Place tick marks (A and B) on a sheet of scrap paper, indicating the separation between two points on the map, in this case the center of the north–south road and the quarry. Step 2: Bring the tick marks to the graphic scale on the map and read the ground distance directly.

Rulers for measuring distances on 7 1/2-minute and 15-minute quads are provided along the margins of these two facing pages.

GRID NORTH

The Universal Transverse Mercator Grid System of military grid zones has been adopted for an area bounded by 80° S and 80° N latitude (Fig. 68). It consists of 60 grid zones, each 6° of longitude in width. Grid declination is the angle between north–south UTM grid lines and true north, the latter of which is indicated by meridians. Quadrangles that employ the UTM System include margin information that describes it.

Note: Section lines do not comprise this grid. This grid is indicated by labeled tick marks along the map's borders.

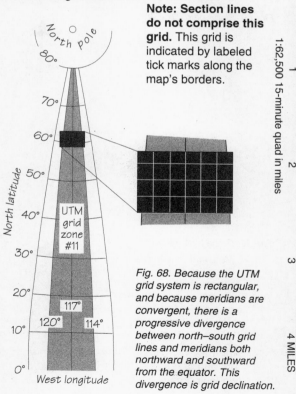

Fig. 68. Because the UTM grid system is rectangular, and because meridians are convergent, there is a progressive divergence between north–south grid lines and meridians both northward and southward from the equator. This divergence is grid declination.

TRIGONOMETRIC SOLUTIONS OF STRATIGRAPHIC THICKNESS

T = stratigraphic thickness
S = slope distance (determined by pacing or taping)
y = slope angle
x = dip angle

Cases where slope and dip are in opposite directions.

Fig. 69. Slope plus dip < 90°.

$T = S \sin(x + y)$

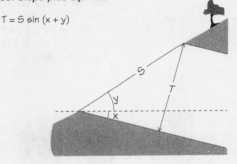

Fig. 70. Slope plus dip > 90°.

$T = S \cos(x + y - 90)$

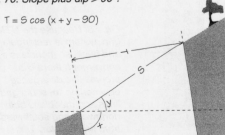

Cases where slope and dip are in the same direction.

Fig. 71. Slope < dip.

$T = S \sin(x - y)$

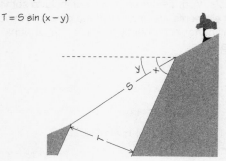

Fig. 72. Slope > dip.

$T = S \sin(y - x)$

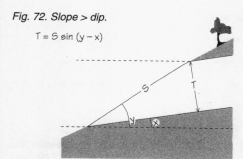

LITHIC PATTERNS AND SYMBOLS

The most inclusive library of lithic symbols is that of the U.S. Geological Survey, and although a number of computer programs offer fill-patterns, only Adobe Illustrator's Gallery includes the entire U.S.G.S. collection. Unhappily, most computer-generated fill-patterns are bit-maps, rather than vector objects, so lines tend to be coarse. Figure 73, which I drew with Macromedia's FreeHand, is an example.

Figure 73 presents an abbreviated scheme of lithic symbols. Your field area might not include some of these elements, yet might include others. Caution: Use fill patterns prudently to avoid a 'chart-junk' appearance.

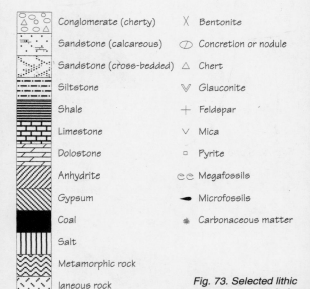

Fig. 73. Selected lithic patterns and symbols.

STRUCTURAL MAP SYMBOLS

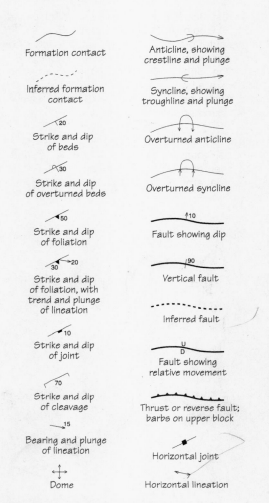

Fig. 74. Selected structural map symbols.